What Is Water?

by Lisa M. Herrington

Children's Press®
An imprint of Scholastic Inc.

Content Consultants
American Geosciences Institute

Copyright © 2023 by Scholastic Inc.

All rights reserved. Published by Children's Press, an imprint of Scholastic Inc., *Publishers since 1920.* SCHOLASTIC, CHILDREN'S PRESS, LEARN ABOUT WATER, and associated logos are trademarks and/or registered trademarks of Scholastic Inc. The publisher does not have any control over and does not assume any responsibility for author or third-party websites or their content.

No part of this publication may be reproduced, stored in a retrieval system, or transmitted in any form or by any means, electronic, mechanical, photocopying, recording, or otherwise, without written permission of the publisher. For information regarding permission, write to Scholastic Inc., Attention: Permissions Department, 557 Broadway, New York, NY 10012.

Library of Congress Cataloging-in-Publication Data Available

978-1-338-83695-0 (library binding) | 978-1-338-83696-7 (paperback)

10 9 8 7 6 5 4 3 2 1   23 24 25 26 27

Printed in China 62
First edition, 2023

Series produced by Spooky Cheetah Press
Book prototype and logo design by Book&Look
Page design by Kathleen Petelinsek, The Design Lab

Photos ©: cover: David Doubilet; 3 background and throughout: Freepik; 8 left: THEPALMER/Getty Images; 15: kali9/Getty Images; 18 globe: NASA; 19: Photographerlondon/Dreamstime; 27: jf/Getty Images; 28 bottom: unclepodger/Getty Images; 30: Courtesy of Stephanie Peltier.

All other photos © Shutterstock.

TABLE OF CONTENTS

The Wonders of Water 4
Chapter 1: Why We Need Water 6
Chapter 2: Water World 12
Chapter 3: Fresh Water 16
Chapter 4: States of Water 22
Activity 28
Water Warrior 30
Glossary 31
Index/About the Author 32

The Wonders of Water

All living things need water to survive. That includes people, plants, and animals. Without water, life on Earth would not exist. Water is the most important **natural resource** on our planet.

We need to drink plenty of water each day to stay healthy.

CHAPTER 1

Why We Need Water

Water is all around us. It is also inside us. Our bodies would not work properly without water. Water helps us digest food. It also keeps our bodies at the right temperature. Sweating is one way your body cools itself. When you sweat, your body releases water. As that water dries, your body is cooled.

Water makes up more than half a person's body weight.

Water is great for drinking. But we also need it for many other things. We use water for cooking and washing. Firefighters use water to put out fires. Farmers use it to grow crops.

Large boats travel on the water to carry goods around the world. The power from rushing water is used to create electricity. People also like to swim and play in water!

Other living things depend on water as well. Water helps plants grow. They take in water through their roots. Animals drink water, too. It also provides **habitats** for many living things. All kinds of plants and animals make their homes in water.

Plants need water to make food in their leaves.

Humpback whales like this one live in every ocean.

There is more water than land on Earth's surface!

CHAPTER 2

Water World

Earth is often called the Blue Planet. Look at the picture of Earth and see if you can guess why. More than 70 percent of Earth's surface is covered in water. There are two kinds of water on Earth—salt water and fresh water.

Almost all Earth's water—97 percent—is salt water. It is found in Earth's five oceans: the Pacific, Atlantic, Arctic, Indian, and Southern. Salt water is also found in seas. Those are large bodies of water mostly surrounded by land. Seas are often parts of oceans.

The shore is where oceans, seas, and other large bodies of water meet the land.

The Pacific is the world's largest and deepest ocean.

Water can be very powerful. As rivers rush along, they can change the shape of the land.

Over millions of years, the Colorado River has carved through surrounding rocks. The rushing water formed the Grand Canyon.

CHAPTER 3

Fresh Water

Of all Earth's water, just a small amount (about 3 percent) is fresh water. That is the kind of water we drink. Some of that fresh water is found in lakes, rivers, and streams. Lakes are bodies of water surrounded by land on all sides. Rivers and streams are moving water. They flow into lakes, seas, and oceans.

Most of Earth's fresh water is locked up in snow and ice. If snow piles up, over time it can form a mass called a **glacier**. A glacier that covers a large area is known as an ice sheet. The ice sheets in Greenland and Antarctica hold most of the world's fresh water. Large pieces of ice that break off glaciers are called icebergs.

Greenland ice sheet

Most glaciers take more than 100 years to form.

There is more water underground than in all of Earth's lakes, rivers, and streams!

Groundwater in an aquifer is usually not visible. We can see the water in this aquifer because the land above it collapsed!

Some of Earth's fresh water is **groundwater**. When it rains or snows, water seeps into soil and fills the cracks and spaces underground. Some of that groundwater makes its way into rivers and streams. Some groundwater can stay in underground layers of sand and gravel, called aquifers. Some groundwater can dissolve the rock underground. That is how caves are formed.

CHAPTER 4

States of Water

Water is the only substance found naturally in three states: liquid, solid, and gas. The state of water changes with temperature.

When water is a liquid, it is wet. The water you drink is liquid. Rain is liquid water that falls from the sky. The water that flows down a river or stream and moves to the ocean is also liquid.

The water we use for watering plants is liquid.

When it is very cold and water freezes, it becomes a solid. That is something hard and firm like an ice cube. Glaciers and icebergs are all frozen water.

Water vapor, a gas, is invisible. It is found in the air. If it cools enough as it rises into the air, it can form small droplets. These small drops of liquid water can form clouds, fog, and steam.

Water changes states all the time. It changes as it moves from oceans, rivers, and lakes to the air. Then it moves back from the clouds to land again. This never-ending journey is known as the water cycle.

Water is found in many forms on Earth. It is an important resource for all living things. That is why we need to take good care of it!

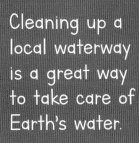
Cleaning up a local waterway is a great way to take care of Earth's water.

The water you drink today is likely the same water that was on Earth when dinosaurs roamed!

ACTIVITY

THE POWER OF WATER

Erosion is the wearing away of land by water, wind, or ice. It changes the shape of the land. This activity shows how erosion happens as water picks up pieces of dirt and moves them along.

YOU WILL NEED

- An adult's help
- Deep container at least 7 inches (18 centimeters) deep by 15 inches (38 centimeters) long
- Sand
- Scissors
- Two paper cups
- Water

STEPS

1. Fill the container halfway with sand.
2. Create a hill so there is more sand on one side of the container.
3. With an adult's help, use the scissors to poke a hole in the bottom of one paper cup.
4. Fill the second cup with water.
5. Hold the cup with the hole over the side of the container with more soil.
6. Pour the water into the cup with the hole.
7. Observe the water as it flows over the sand. What happens? Can you tell where the water travels?

WATER WARRIOR

MEET AUTUMN PELTIER

Autumn is a teenager who lives in Canada. She is a member of the Wiikwemkoong First Nation. Some Indigenous communities in Canada do not have clean drinking water because it is polluted with harmful waste. They have to boil the water to make sure it is clean before they drink it.

Autumn has spoken out about this problem. In part because of her work, the Canadian government has been working to bring clean water to First Nations communities. Autumn encourages other kids to take action and use their voices to protect water.

GLOSSARY

glacier (GLAY-shur) a slow-moving mass of ice found in mountain valleys or polar regions

groundwater (GROUND-waw-tur) water below the ground that can be used for drinking and other purposes when wells are dug into it

habitats (HAB-i-tats) places where plants or animals are usually found

natural resource (NACH-ur-uhl REE-sors) something that is found in nature and is valuable to humans, such as water or trees

water vapor (WAW-tur VAY-pur) water in the form of a gas

INDEX

aquifer **20, 21**
cave **21**
Earth **4, 12–14, 17, 18, 20, 21, 25–27**
erosion **28, 29**
fresh water **13, 17, 18, 21**
glacier **18, 19, 25**
groundwater **20, 21**

iceberg **18, 25**
ice sheet **18, 25**
lake **17, 20, 26**
liquid **22–25**
ocean **14, 15, 17, 26**
plant **4, 10, 23**
rain **21, 22**
river **16, 17, 20–22, 26**

salt water **13, 14**
sea **14, 17**
snow **18, 21**
solid **22–25**
stream **17, 20–22**
water cycle **26**
water vapor **24, 25**

ABOUT THE AUTHOR

Lisa M. Herrington has written hundreds of books and articles for kids over the years. When Lisa isn't writing and it is hot outside, she enjoys swimming in the pool with her family.